SPRINGVALLEY ELEMENTARY
RESOURCE CENTRE

DESIGNS IN SCIENCE
TECHNOLOGY IN ACTION

SALLY and ADRIAN MORGAN

Technology in Action
Copyright © 1994 by Evans Brothers Limited
All rights reserved. No part of this book may be reproduced or utilized in any form or by any means, electronic or mechanical, including photocopying, recording, or by any information storage or retrieval systems, without permission in writing from the publisher. For information contact:

Facts On File, Inc.
460 Park Avenue South
New York NY 10016

Library of Congress Cataloging-in-Publication Data
Morgan, Sally.
    Technology in Action / Sally and Adrian Morgan.
       p. cm. — (Designs in science)
    Includes index.
    ISBN 0-8160-3126-6 (acid-free paper)
    1. Biotechnology — Juvenile literature. 2. Biology — Juvenile literature. 3.Engineering design — Juvenile literature. I. Morgan, Adrian. II. Title. III. Series: Morgan, Sally. Designs in science.
GB662.3.M67   1994
660'.6—dc20   93-7770

Facts On File books are available at special discounts when purchased in bulk quantities for businesses, associations, institutions or sales promotions. Please call our Special Sales Department in New York at 212/683-2244 or 800/322-8755.

10 9 8 7 6 5 4 3 2 1

This book is printed on acid-free paper.

Managing Editor: Su Swallow
Editors: Catherine Bradley and Kate Scarborough
Designer: Neil Sayer
Production: Jenny Mulvanny
Illustrations: Hardlines, Charlbury
                  David McAllister

Acknowledgments

For permission to reproduce copyright material the authors and publishers gratefully acknowledge the following:

Cover (main photograph) JC Revy, Science Photo Library (inset) Hank Morgan, Science Photo Library
Title page Brian Brake, Science Photo Library
Contents page Sally Morgan, Ecoscene **page 6** (top) Whittle, Ecoscene (middle) Robert Harding Picture Library (bottom) JAL Cooke, Oxford Scientific Films **page 7** Caroline Cooper, Ecoscene **page 8** (top) Andy Purcell, Bruce Coleman Ltd (bottom) Robert Harding Picture Library **page 9** Norbert Wu, NHPA **page 10** (top) Ecoscene (bottom) Mark Hamblin, Oxford Scientific Films **page 11** Ecoscene **page 12** (top) ZEFA (bottom left and right) Ecoscene **page 13** ZEFA **page 14** Flip Chalfant, The Image Bank **page 15** David Parker, Science Photo Library **page 16** Jules Cowan, Bruce Coleman Ltd **page 17** Eric Crichton, Bruce Coleman Ltd **page 18** (top) Ralph and Daphne Keller, NHPA (bottom) Laurie Campbell, NHPA **page 19** (top) Jules Cowan, Bruce Coleman Ltd (bottom) ZEFA **page 20** (top) Dr Norman Myers, Bruce Coleman Ltd (middle) Hans Reinhard, Bruce Coleman Ltd (bottom) Cooper, Ecoscene **page 21** (top) ZEFA (bottom) Science Photo Library **page 22** Carl Schmidt-Luchs, Science Photo Library **page 23** (top) ZEFA (bottom) Ecoscene **page 24** (left) Towse, Ecoscene (right) David Scharf, Science Photo Library **page 25** Sally Morgan, Ecoscene **page 26** Sally Morgan, Ecoscene **page 27** (top) Ecoscene (bottom) GH Thompson, Oxford Scientific Films **page 28** (top) David Parker, Science Photo Library (bottom) Sally Morgan, Ecoscene **page 29** (top) Robert Harding Picture Library (bottom) Hattie Young, Science Photo Library **page 30** (top) Sally Morgan, Ecoscene (bottom) Eric Soder, NHPA **page 31** Dr Norman Myers, Bruce Coleman Ltd **page 32** (top) Williams, Ecoscene (middle) Robert Harding Picture Library (bottom) Jane Burton, Bruce Coleman Ltd **page 33** (top) WS Paton, Bruce Coleman Ltd (bottom) Kim Taylor, Bruce Coleman Ltd **page 34** (top) ZEFA (bottom) Andy Purcell, Bruce Coleman Ltd **page 35** (top) Sue Ford, Ecoscene (bottom) Science Photo Library **page 36** (top) Adrian Davies, Bruce Coleman Ltd (bottom) Jane Burton, Bruce Coleman Ltd **page 37** Stephen Dalton, NHPA **page 39** (top) ZEFA (bottom) Ecoscene **page 40** (top) Robert Harding Picture Library (middle) Scott Camazine, Science Photo Library (bottom) Robert Harding Picture Library **page 41** Sally Morgan, Ecoscene **page 42** Ecoscene **page 43** (top) US Department of Energy/Science Photo Library (bottom) Robert Harding Picture Library **page 44** (top) Philippe Plailly, Science Photo Library (bottom) Martin Bond, Science Photo Library **page 45** JC Revy, Science Photo Library

DESIGNS IN SCIENCE
TECHNOLOGY IN ACTION

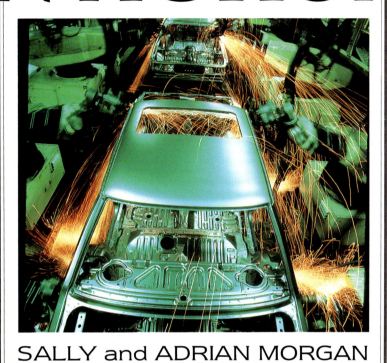

SALLY and ADRIAN MORGAN

Facts On File

NOTE ON MEASUREMENTS:

In this book, we have provided U.S. equivalents for metric measurements when appropriate for readers who are more familiar with these units. However, as most scientific formulas are calculated in metric units, metric units are given first and are used alone in formulas.

Measurement

These abbreviations are used in this book.

METRIC		**U.S. EQUIVALENT**	
Units of length			
km	kilometer	mi.	mile
m	meter	yd.	yard
cm	centimeter	ft. or '	foot
mm	millimeter	in. or "	inch
Units of temperature			
°C	degrees Celsius	°F	degrees Fahrenheit

Technology in Action is one book in the Designs in Science series. The series is designed to develop young people's knowledge and understanding of the basic principles of movement, structures, energy, light, sound, materials, and water, using an integrated science approach. A central theme running through the series is the close link between design in the natural world and design in modern technology.

Contents

Introduction 6

Design and evolution 8
  Evolution
  Survival of the best designs
  Blueprints

Artificial selection 16
  Plant breeding
  Animal breeding
  Genetic engineering

Biotechnology 24
  Enzymes
  Biosensors
  Cleaning up pollution

Defense mechanisms 32
  Pesticides
  Natural pesticides
  Antibiotics

Control systems 38
  A natural control center
  Microchips and computers
  Feedback mechanisms
  Artificial intelligence
  Robots

The future 44

Glossary 46

Index 47

6 DESIGNS IN SCIENCE

Introduction

If you were to look at all the different plants and animals in the world, you would see an incredible variation in size and shape. Each individual living plant or animal is the result of a slow process of natural design called evolution. Evolution is the gradual change in the characteristics of species of plants and animals. Charles Darwin proposed his theory of evolution more than 100 years ago and most of his ideas are still considered valid today. He described evolution as a process whereby the organisms best suited to a particular environment would be most likely to survive and reproduce. In doing so, their characteristics would be passed on to the next generation. He called this process "natural selection." It is also known as the survival of the fittest.

Both this lizard and the pipelines are examples of successful designs. The lizard is perfectly camouflaged among the leaves, while the pipeline allows large volumes of oil to be transported efficiently.

In fossil records, many strange looking organisms have been discovered. Some of them do not resemble any modern organisms at all. For example, there was an armored worm, just a couple inches long, that lived more than 500 million years ago. It had several conical shells arranged symmetrically around its body. Scientists think the worm used its armored shells as protection against predators. However, this design cannot have been that successful because this feature is not present in modern day worms. These unusual designs can be thought of as prototypes or trials that did not work properly and were not suited to the environment. In the process of natural selection, these organisms were not successful, so they died out. The organisms that did reproduce and were able to survive until the modern day are the successful variations of these prototypes.

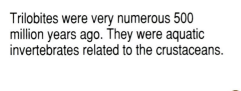

Trilobites were very numerous 500 million years ago. They were aquatic invertebrates related to the crustaceans.

Human designers carry out a similar process when designing a new product or building. There may be many ideas, and several prototypes may be tested, but only those that meet the necessary standards and requirements

TECHNOLOGY IN ACTION | 7

There may be more than 10 million different species of living organisms on this planet, although only 1.4 million have been identified.

The giraffe has evolved a long neck that enables it to reach the higher branches of trees that other herbivores are unable to reach.

Key words
Evolution the gradual change in the characteristics of an organism.
Natural selection theory that those species best suited to an environment will live to reproduce, passing on their characteristics.
Prototype a pre-production or test version of a product.
Technology the practical application of science in everyday life.

survive to go into production. This is also a form of selection. Today, many designers are looking to the natural world, both for ideas and for ways of testing new products. There are many interesting parallels between the natural world and that which we construct for ourselves. For example, many animals have jaws that resemble a pair of pliers and work on the same principle. In another example, engineers designing a pipeline oil distribution network discovered that the computer-generated answer to the problem was very similar to the layout of blood vessels in the mammalian gut. Mammals have to move blood around the gut in the same way that oil had to be moved along the pipeline. Natural selection had evolved an ideal solution to the problem, producing a very efficient distribution network.

Technology plays an increasingly important role in our everyday lives. In this book we will examine some of the features of the natural world that scientists are studying in their search for answers to some of our modern technological problems. An ever-increasing number of manufactured products such as detergents, antibiotics, and pesticides contain natural substances that enable them to work more efficiently. We will examine how nature and people use these substances and how the latest developments may improve the quality of our lives in the future.

Important words are explained at the end of each section under the heading **Key Words** and in the glossary on page 46. You will find some amazing facts in each section, together with some experiments and some questions for you to think about.

8 DESIGNS IN SCIENCE

Design and evolution

The peppered moth occurs in two forms, a normal peppered form and a black form, which has adapted to life on polluted trees.

Both natural evolution and human design processes involve selection. Those designs that are best suited to their purpose in their particular environment are chosen. For example, if the environment changes, the animals or plants living in that environment have to adapt or change in order to survive. Since the beginning of the industrial revolution, the environment has been badly affected by pollution. This has affected many animals and plants, such as the peppered moth. This black, gray and white moth is perfectly camouflaged when it rests on tree trunks. Occasionally it occurs in an all-black form. As a result of industrialization, the environment changed. Soot from chimneys darkened the trunks of trees in industrial areas. The black form was ideally camouflaged when resting on these darkened tree trunks, while the mottled moths were easily spotted by birds. The black form became very common in the polluted areas, while the mottled form survived in the rural areas. This example shows how a change in the environment can affect the survival of a species such as the peppered moth.

Evolution

If you were to study all the individuals of a population, you would find that there were many small differences. Individuals of the same species always show some variation. For example, not all dogs of the same breed look identical. All the children in your class at school will have slightly different colored hair or eyes, or will be a different height. It is variation that makes us all different, even different from our parents. This variation is

Human beings all belong to the same species. However, as this picture shows, each individual is different from the next.

TECHNOLOGY IN ACTION 9

 How many differences can you spot between your face and the faces of a friend, a brother or a sister?

essential to the process of evolution. Darwin's theory of natural selection means that the best variations will survive to reproduce. Over a long period of time, the species' characteristics will adapt to suit the environment better. This process of adaptation is happening all the time, because the environment itself changes continually. So some organisms have advantages over others at different times.

Continual evolution has been the driving force for the great variety of species around today. Every element of every species works because of evolution. But it is difficult to understand how a complex organ such as an eye could slowly evolve. It is possible, however, to theorize on how this might have occurred. The eye is a sense organ that can detect light and send messages about that light to the brain. We know from the fossil records that the eye did not suddenly appear. It evolved slowly, over many millions of years, into the various forms that we know today. Many millions of years ago there was probably an animal that had a few light-sensitive cells on the surface of its skin, or epidermis. This animal would have an advantage over the other animals that did not have these cells, for it would be able to detect changes in light level, using this information to find food more efficiently. Over a period of time, a few variations appeared. Perhaps some of the light cells were arranged at the bottom of a slight depression. This is better at trapping the light, so the depression may have eventually deepened to form a pit. Later, a substance such as a jelly-like mucus may have been trapped in the pit, forming a poor but workable lens. Given time, we can see how a primitive eye could have evolved. The basic principle is that even a slight advantage can lead to significant change. Think how much easier life would be if you had just five percent of your vision, compared with none at all.

Eyes have evolved into many different forms. Above is a squid's eye, which is surprisingly similar to the eye of a human (below). They developed separately but prove that the design is successful.

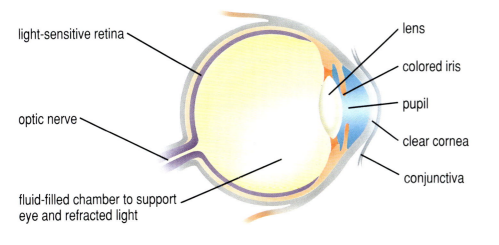

The appendix does not have much of a role in the body of a modern human being, although it helps herbivorous animals such as rabbits to digest their food. In time it may disappear from the human body completely.

10 DESIGNS IN SCIENCE

EXPERIMENT

Investigating variation in hands

There is considerable variation among just a few people of the same age. If you look at your friends in your class at school you will see many differences, such as the color of hair and eyes, height and even the length of fingers and shoe size. In this experiment you will make a survey of the length of the longest finger in a group of people, perhaps in your class at school. You will discover how much variation there is between people. You will need a ruler, a notebook, pencils and some graph paper.

1 You will need to measure the length of the middle finger of the right hand using people of a similar age. Ask the person to bend their finger at the knuckle while keeping their fingers straight. Measure from the edge of the knuckle to the tip of the finger. Do not include the length of nail that projects beyond the finger tip. Record the measurement in your notebook. You will need to measure the middle finger of at least 20 people.

2 Look at your results. Group your finger lengths into 0.5 cm (0.2") categories, for example 7.0-7.5 cm, 7.5-8.0 cm (2.8"–3", 3"–3.2") and so on. Record the number of lengths that fall into each category.

3 You can now plot your results in a bar chart, as shown below.

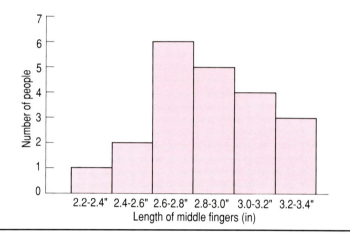

Survival of the best designs

The spines of the hedgehog are hollow, so they are very light, and because of their design they are also very strong.

The process of natural selection, which ensures that organisms are well adapted to the environment, continually refines and improves upon the design of the organism. The word *environment* refers to the surroundings of an organism and includes all the different factors that may affect its survival. Such factors will include other living organisms, the soil, the climate, and the supply of water and oxygen. Those organisms that are best adapted to a particular environment will be most likely to survive and produce offspring. Their genes are passed on to the next generation. A good design will make it easier for an animal to find food or avoid predators. Often it is a combination of characteristics that produces a successful design. This design will be constantly updated and improved until the optimum design is reached. If the environment changes, the design will have to change as well, or the creature may not survive.

The ideal design is the one that will give the best chance to the organism of surviving long enough to pass on its genes to its offspring.

TECHNOLOGY IN ACTION 11

The upturned bill of the avocet is ideal for filtering water to gather up insects that may be floating on the surface.

A completely different design of bill for a completely different use. Oystercatchers pry shells open with their bills.

However, although it is often almost impossible to judge how good a particular design is, it is sometimes possible to measure the degree of success of one particular part of a design. For example, the bill of a bird has to be designed so that the bird can feed efficiently. There are many designs of bill, each adapted to a different style of feeding. Many birds such as the gulls and crows have all-purpose bills, which are able to feed on a wide range of foods. Other birds, however, have specialized bills that are designed to feed on one type or source of food. For example, the curved bill of the avocet is designed to filter food. It can skim insects from the surface of mud or from shallow water. Some birds such as the herons and egrets have pointed, dagger-like bills, which are ideal for catching fish.

The design of a structure has to be successful too. For example, a bird has to be light in weight but at the same time must have bones that can resist the force of landing. Evolution has provided birds with hollow bones that have the right balance between weight and strength. This hollow shape, or tube, is seen in many different structures such as the stems of grasses and reeds and even the spines of some animals. However, a tube is not so strong when it is pushed from the side. You can test this for yourself by comparing the resistance of the sides of a cardboard toilet paper roll compared with its end to end strength. The difference is very marked. Engineers can strengthen a hollow shape by adding internal struts or ridges called stringers.

There are other ways of measuring design success. One such way is to consider and compare the costs of a design, such as the energy costs in the manufacturing process or the expense of the materials. Animals such as insects have to balance the cost in energy of finding food with the benefits gained by eating it. Bees, for example, which feed on the pollen and nectar found in flowers, have to consider whether it is more efficient to spend a lot of time at one

12 DESIGNS IN SCIENCE

! *A mushroom may release 16 billion spores (similar to seeds) to ensure that just a few survive and grow to maturity.*

This bee has to work out the economics of collecting pollen.

flower and suck up all the available nectar or move from flower to flower just taking a little. In fact, researchers studying the behavior of some bees in North America have found that these particular bees suck only some of the nectar from the flowers, leaving the rest for later. During the morning, when they need to build up their energy, they fly from flower to flower taking the nectar that is most readily available. Only later on do they go back to the flowers to collect the rest. The flower also gains from this arrangement. Because as the insects visit the plant at least twice, there is a greater chance of pollination.

Another measure of success might be the reproductive ability of the organism. This might be judged by the number of seeds produced or the number of offspring raised. Plants are unable to move around and so their choice of reproduction strategy is important. Should it, perhaps, produce flowers and seeds all through the season or should it grow for longer, building up more food reserves and produce all of its flowers and seeds at the end of the season? The wheat plant, for example, produces large leaves during the summer months using the food made in photosynthesis. Finally, at the end of summer, the seed head is produced. From this time on, no more leaves are produced and all the plant's energy goes into seed production.

! *The coconut (below) provides a fruit that floats so it can be carried by tides and currents to remote islands where the seed within the fruit germinates and colonizes new land.*

The sycamore tree produces a lot of seeds to make sure that at least some will have the chance to grow into trees. They are designed so that they spin away from the parent plant.

TECHNOLOGY IN ACTION | 13

The landrover is designed to travel over rugged country. It is able to move heavy loads up steep slopes and through water.

 What features of your family car do you like? What features would you like to change? Could the car be better suited to your family's needs?

As a result of growing all summer, the wheat has the resources to produce large seeds. A large seed with its own food store gives the new plant a good start when the seed germinates.

It is somewhat easier to measure the success of artificial designs produced by people. A good design may be measured by how much energy it uses, how safe it is or, perhaps, how long it lasts. However, everybody has their own ideas of whether or not a particular product is successful. In car designs, engineers look at many different features. They consider hundreds of factors, including how much gasoline the car will use, how reliable the engine is, and how safe the car is in an accident. People choose to buy a certain design of car to suit their lifestyle. A family may prefer a large, economical car with lots of space for luggage whereas a single person may want a fast car and have little need for luggage space.

The success of a design can also be measured by the number of sales. A well designed product will make lots of sales. There will be good reports about the product and people will want to buy it. An unattractive or unreliable design will not sell well, and the manufacturer will try either to improve the design or design something completely new. In this way design success is measured by the sales record.

The consumer is also helped by certain standards that are applied to products. Countries often have certain standards that a product must meet. For example, upholstery has to be made of

14 DESIGNS IN SCIENCE

fabric that is fire resistant so that the fabric will not burn easily. If the product reaches the right level of quality, it can be awarded the standard. This standard is usually indicated by a symbol marked on the product.

Today people are very concerned about the way we are damaging the world environment, and in some countries there are schemes that measure the environmental damage that a product might do, or damage that is incurred during the manufacturing process. A product with a low environmental impact will be awarded a special green symbol, helping the consumer decide which product to buy.

Blueprints

When a designer or architect designs anything, the design is drawn as a series of plans. These are often referred to as blueprints, because, in the past, plans were copied using a process that produced a blue-tinged paper. A blueprint shows where all the building materials are placed and how the connections are made.

In a living organism the blueprint is in the form of genetic information held within the nucleus of a cell.

Each organism has a particular number of chromosomes present within the nucleus of its cells. For example, a human being has 46 chromosomes, which occur in 23 pairs, while the tiny fruit fly has just four pairs. A chromosome is a very long thread-like structure made of the chemical DNA (deoxyribonucleic acid). The chromosome can be subdivided into specific

The people in this photo are architects. They are drawing up plans for a new building. Everything that has been made by people has been designed.

TECHNOLOGY IN ACTION | 15

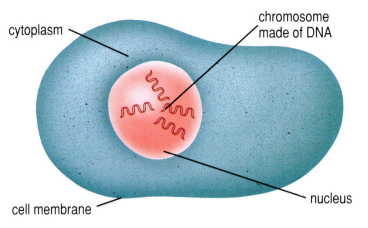

A typical animal cell, showing its nucleus, which contains chromosomes. These are the "blueprints" of the living organism.

Computers are often used to produce blueprints, in this case, the layout of a ventilation system.

lengths called genes, in the same way that a chain of beads is made up of separate beads. Each gene is responsible for a particular characteristic. For example, there are genes present that determine the color of your hair, your eyes and even your ability to roll your tongue or taste particular chemicals. Each individual receives chromosomes from each of their parents. In humans, the mother and father each give the offspring 23 chromosomes, so their children get a complete set of 46 chromosomes. Any change that occurs in the makeup of the parents' chromosomes can be passed on to the offspring. Changes to the genetic material are called mutations. Sometimes a mutation can be harmful and the individual may not survive. In this way, the harmful mutation dies out with the individual. However, other mutations may be advantageous, making the individual more likely to survive and pass on the mutation to its own offspring. Mutation, therefore, has a very important role to play in evolution.

Designers produce blueprints too. These detailed plans show the arrangement of all of the parts of a structure. For large and complex structures, there will be many different plans, each showing a particular aspect of the overall design. For example, the set of plans for a building may have individual plans for room layouts, electrical circuits, position of pipes and so on. These can be very useful, for example, during an emergency such as a fire, when the emergency services will use the blueprints to help them plan the rescue.

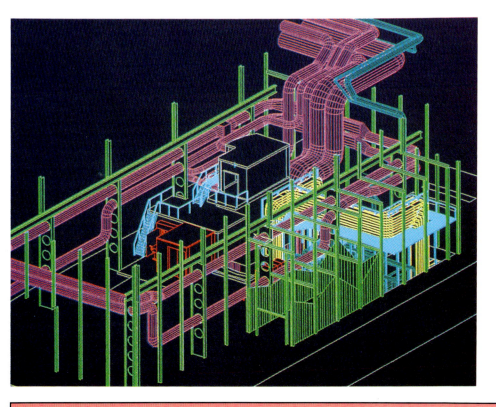

 The ability to roll the tongue is controlled by a gene. Can you roll your tongue? Can your parents roll their tongues?

Key words
Adaptation the process of adjustment of an organism to environmental conditions.
Mutation a sudden change in genetic information from generation to generation.
Selection making a choice.
Variation small differences, a change from that which is normal.

16 DESIGNS IN SCIENCE

Artificial selection

The processes of natural selection among plants and domestic animals can be carried out artificially. Instead of allowing the environment to select the best suited variation, the selection can be carried out by people. This involves careful choice of parents and then careful selection of the offspring. Over several generations, changes can be seen.

Plant breeding

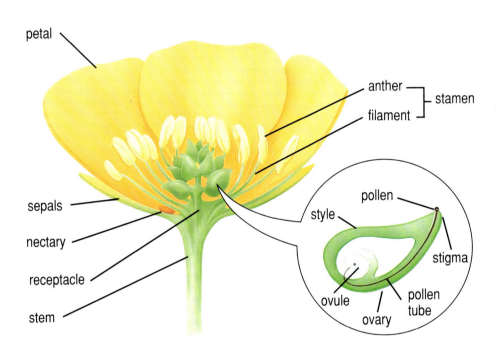

Some plants reproduce by using flowers. A flower consists of many parts, including the petals, sepals and, of course, the male and female reproductive structures. Most flowers contain both male and female structures. The male part, called the anther, produces grains of pollen that contain a male nucleus. The female part, the carpel, consists of a stigma, style and ovary. The ovary contains an ovule, inside which is the female egg cell. In the process of pollination the pollen grain is moved from the anther to the stigma. This may be done by an animal such as a bee or by the wind or even by water. The pollen grain then produces a tube, which grows down the style into the ovary and then the ovule. There the male nucleus will fertilize the female egg cell. The result is an embryo protected within a seed. The embryo will develop into a new plant.

In self-pollination the pollen pollinates a stigma on the same individual plant, possibly in the same flower. However, this produces in-breeding and results in less variation. This is not ideal and cross-pollination is preferred. Most flowers have special mechanisms to prevent self-pollination. For example, the pollen may be produced before the stigma is ready to receive the pollen, so that the pollen can only pollinate another flower. Some plants have single-sexed flowers to prevent self-pollination.

Cross-sections of flowers showing the reproductive parts. Above is the buttercup with its anthers higher than its stigma. Below, the tulip flower has a large, central stigma surrounded by anthers.

TECHNOLOGY IN ACTION 17

During artificial pollination the anther with pollen is rubbed on a stigma of another flower.

? *How do plants get pollinated naturally?*

The orchid in the middle of this photograph has been produced by cross-pollinating the two other orchids.

Modern-day crop and garden plants have been bred from wild species of plants. These plants are now very different from their wild ancestors, bearing large and colorful flowers or heavier seeds. The traditional way of producing a new variety of plant is to cross-breed two plants of the same species that have the desired characteristics. It is possible to control the pollination of flowers to carry out a cross. The pollen of the selected parent is removed from the anthers, often using a soft brush, and carefully wiped onto the stigma of the other parent. The flower receiving the pollen has to be covered so that no other pollen can reach the stigma either before or after the artificial pollination. The seeds resulting from the cross are collected and germinated. The plant breeder will carefully examine the new plants and will select only those with the desired characteristics for further breeding.

It may take many generations and thousands of plants to produce a plant with the desired characteristics because plant breeding is not a simple process. Sexual reproduction creates variations and many of these will produce worse features than those of the original parent plants. For example, a breeder may succeed in producing a plant of the desired color, but the flower may have a strange shape or the plant may lose its resistance to disease. The breeder either throws away the new plant and starts again, or tries to breed additional characteristics into the new plant while preserving its color.

Farmers have been carrying out artificial selection for more than five thousand years, since agriculture first developed in the

18 DESIGNS IN SCIENCE

Cultivated wheat contains 42 chromosones while wild wheat has only 14 chromosones.

Cultivated wheat has a sturdy stem and large seedhead that does not shatter easily (right) compared to wild grasses, which have smaller seedheads and wispy stems.

Middle East. Early farmers would collect the seedheads of wild grasses and sow them the following year. They tended to pick the large, more obvious seedheads, so slowly the seedhead of the planted grasses increased in size. These semi-wild grasses were sown in places they would not normally grow and so came into contact with other species of wild grass. This resulted in crosses that produced new species of grass. Each year, the farmers would collect those seedheads that had the largest seeds, and plant them the following season. Several thousand years later, modern cereal plants such as wheat and barley, which have large seedheads, are the result.

Scientists are still trying to improve the various strains of modern wheat. They carefully cross one variety with another and select the best variations. For example, they are trying to breed a wheat that has a large seedhead with a shorter stem so it does not collapse under heavy rain. They are also attempting to improve the natural resistance of the wheat to some of the diseases that affect cereals.

However, plant scientists have to be careful that they do not get rid of all variation. Visually, the wheat plants growing in a field all seem to be identical to one another. They have, in fact, been so carefully bred that much of the variation that used to be present has been lost. This can cause problems, for if all the plants are equally susceptible to a certain disease, it can wipe out an entire crop. For this reason, scientists are developing seed banks in which seeds of older varieties of wheat and wild grass are stored. The wild grasses may be important in the future for providing new sources of variation or new genes that give resistance to particular diseases. There are a great number of wild relatives to our modern crop plants. There are also many other species of plant that could be used as crops in the future.

Rain forests are the most diverse habitat in the world, with more species of plants and animals per acre than any other habitat. It is estimated that at least 50 percent of all the species

The modern tomato plant contains genes that have come from at least nine wild relatives.

TECHNOLOGY IN ACTION — 19

Genetic uniformity in crop plants might have played a role in history! The Mayan civilization of Mexico may have collapsed because their staple food, maize, lacked disease resistance and the crops failed.

Why is it important to conserve the world's remaining rain forests?

The rain forest (top) has hundreds of different species of plant. There is enormous genetic variety, whereas the sunflowers in a field are all almost exactly the same.

on the planet are found in the rain forests and many species are still unknown to science. One of the concerns about deforestation in the rain forests is the loss of the different animals and plant species. When rain forests are destroyed, undiscovered species are lost too. It is important to save as much of the rain forest as possible, since the rain forests are huge, natural seed banks.

DESIGNS IN SCIENCE

Animal breeding

Domesticated animals have also undergone a process of artificial selection. The modern dairy cow is a perfect example. Much larger than the dairy cow of a hundred years ago, it is capable of producing a considerable amount of milk each day. It is true to say that, over time, farmers have gradually redesigned the cow. The pig has also undergone many changes. The modern pig is many times larger and heavier than the wild pig.

Breeding animals such as cows is a lengthy process. It takes several years for a calf to mature sufficiently to be able to bear a calf of its own. In addition, since a female cow can produce only one calf at a time, there are a limited number of variations to choose. However, it is possible to use one bull as the father of many different calves. The sperm from a bull with the desired characteristics can be stored and later used to fertilize many cows. Today, the sperm from top class bulls is flown around the world to fertilize cows in different countries.

By selectively breeding animals, it has become possible to incorporate specific design requirements. For example, people today are very aware of the amount of fat in food. Fat can cause

The modern domestic pig (right) is a much heavier animal than the wild bush pig (above).

Holstein cows have been carefully bred so that they produce a lot of milk.

! *Some Holstein cows produce more than 9000 liters (2,340 gal.) of milk per year.*

TECHNOLOGY IN ACTION

Although the many breeds of dog look very different they still belong to the same species.

heart disease by blocking arteries, particularly the coronary arteries near the heart, which can lead to heart attacks. The latest generations of cows are being bred so that their meat contains more protein and less fat.

It seems quite remarkable that the many different breeds of dog that exist today all belong to the same species. The modern dog ranges in size from the tiny Chihuahua to the huge Great Pyrenees. The proof that the dog is still one species is the fact that dogs can all interbreed, to create crosses. Pedigree dogs are those that share the same particular appearance and characteristics. They have been selectively bred to show less variation and so they all look very similar. Breeders of dogs have been refining the pedigree breeds very carefully over the last two centuries.

Genetic engineering

Genetic engineering is a complex process that involves the alteration of genetic material. Genes are responsible for controlling a particular characteristic, for example the manufacture of a protein, the color of the eyes, the gender of the organism, the number of limbs and so on. A large, complicated organism such as a mammal has many millions of genes.

The artificial selection processes take a lot of time and may not be successful. Genetic engineering, on the other hand, can produce a new organism (but very simple) within a few days. If enough detail is known about its genetic structure, the new organism can almost be made to order. The genes of an organism such as a bacterium are quite easy to study, for bacteria have no nucleus, unlike the cells of an animal cell (see page 15). Instead, the DNA lies in the cytoplasm of the cell, so it is easier to access. The length of DNA, and the number of genes, is much shorter in bacteria.

Genetic engineering involves removing a gene from the DNA of one organism and inserting it into the DNA of another organism. This means that the recipient of the new gene will have new capabilities; it can perhaps make a product or do something that it could not do before.

A very good example of the use of genetic engineering is the production of insulin. Insulin is a hormone, or chemical messenger, made from protein. It is produced naturally by mammals to control the amount of glucose (a type of sugar) in the blood. It is normally made in the pancreas and travels in the blood to the liver where it reduces the amount of glucose in the blood after a meal. If the pancreas stops producing insulin, the individual cannot control his or her blood glucose level. People who suffer this condition are

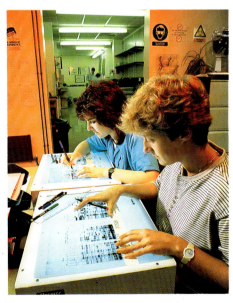

Scientists can take X rays of bands of DNA from an individual. The bands are the patterns of genes that make up the organisms.

22 DESIGNS IN SCIENCE

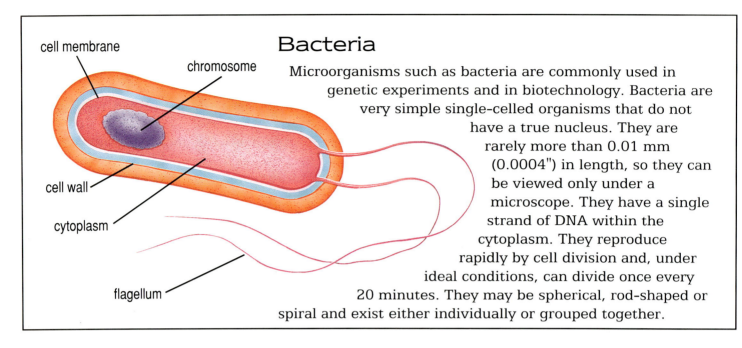

Bacteria

Microorganisms such as bacteria are commonly used in genetic experiments and in biotechnology. Bacteria are very simple single-celled organisms that do not have a true nucleus. They are rarely more than 0.01 mm (0.0004") in length, so they can be viewed only under a microscope. They have a single strand of DNA within the cytoplasm. They reproduce rapidly by cell division and, under ideal conditions, can divide once every 20 minutes. They may be spherical, rod-shaped or spiral and exist either individually or grouped together.

 Do you think it is safe to release genetically altered bacteria into the environment?

called diabetics. In the past, many diabetics were treated with regular injections of insulin that had been extracted from another mammal, usually a cow or pig. However, such insulin is not identical to human insulin, so diabetics suffered from unpleasant side effects. Now scientists have been able to insert the gene for making human insulin into the DNA of a bacterium (see above), enabling it to make human insulin. The insulin can be extracted and used to treat diabetes.

Plants can also be altered so that they produce substances that are required by people. Since they can be grown over large areas, a considerable quantity of the product can be made. A good example of this technology in action involves the tobacco plant. This plant is often infected with a virus called the tobacco mosaic virus, which makes the leaves of the plant become speckled. The virus infects a leaf and, once inside the leaf, it reproduces and spreads throughout the plant. This virus has now been genetically altered so that the virus instructs the plant cells to make a new and quite different protein. There are now fields of tobacco plants in the U.S. that, instead of being made into cigarettes or chewing tobacco, are being used to make medicines that would be too expensive to

The tobacco plant has been used in several genetic engineering trials.

TECHNOLOGY IN ACTION | 23

The biologist is looking at plant chromosomes under a microscope.

 In the future, genetically engineered cows may be able to produce medicines in their milk. Genetically altered sheep may have wool that will drop off when it reaches a certain length.

Oil seed rape yields a lot of oil. In the future scientists may be able to make plastic from the oil it supplies.

 Bacteria can double their numbers every 20 minutes. If you started with just one bacterium, how many would be present in four hours if none died?

Key words
Bacterium a simple single-celled microorganism.
Genetic engineering the alteration of the genetic composition of an organism.
Virus an acellular parasite.

produce in other ways. Researchers are already growing tobacco plants that will yield hemoglobin, the red pigment in blood that carries oxygen to the cells. This can be used to make artificial blood. A small cut is made in the surface of the leaf and the altered virus is inserted into the cut using a dropper. The virus quickly infects the plant and within two weeks all the leaves are making the desired protein.

Genetic engineering may also provide us with new sources of raw materials such as oil. A great range of materials, including many plastics, are made using oil. As world oil reserves dwindle, it is important to look for new ways of making plastics. One such way is to use plants. The mustard plant produces a wide range of natural oils. For this reason, scientists are using genetic engineering methods to insert a gene into the mustard plant to allow it to make plastic from its own oils. The new material is called polyhydroxybutyrate or PHB. However, the manufacture of plastics in this way currently costs up to 20 times more than the cost of making plastic from crude oil. These costs might be reduced by using plants, such as the potato, that would produce a greater yield. The potato plant would make the plastic and store it in its tubers. The tubers could be dug up and the plastic extracted. Already ICI, a British company, is using a plant-based plastic to make biodegradable shampoo bottles. They call this new genetically engineered plastic "Biopol."

24 DESIGNS IN SCIENCE

? *Can you name three foods (including drinks) that make use of yeast?*

Biotechnology

The study of living organisms and developments in genetic engineering can have a significant effect on our daily lives. New natural substances are being discovered all the time and many can be used in industrial processes, replacing artificial substances. Genetic engineering produces new organisms that can be used in industrial processes. The use of biology in industrial processes is called biotechnology.

Biotechnology has already affected our lives. For example, for the last two thousand years people have used natural organisms in their food and drink, namely yeast. Yeast is a microscopic single-celled fungus that is used in bread and beer making and many other processes besides. The brewing and baking industries are dependent on the action of yeast. A better biological understanding of yeast leads to better brewing and baking.

Today, there are two organisms that are used extensively in biotechnology; these are bacteria and fungi. Bacteria are simple, single-celled organisms that are easy to grow and that reproduce rapidly (see page 22). Fungi, a group of organisms that includes mushrooms and toadstools, are in fact a special

Both the shaggy ink cap (below) and the tiny yeast cells (below right) belong to a group of organisms called fungi.

TECHNOLOGY IN ACTION | 25

EXPERIMENT

Growing yeast

Yeast is very important in bread-making since it produces carbon dioxide gas, which causes the bread to rise and become light and airy in texture. It is also used in making beer and wine where the yeast produces alcohol. In this experiment you will investigate the effect of temperature and sugar on the growth of yeast. You will need a large packet of dried yeast, some sugar, three jam jars, some water and a set of scales.

1 Place one large teaspoon of dried yeast into a jam jar and add two teaspoons of sugar. Add about 3½ ounces of water and stir.
2 Add the same amount of yeast to the second jam jar but add two teaspoons of yeast. Add the same amount of water and stir as before.
3 Leave both jam jars in a warm place for 30 minutes. Then observe the differences.

4 Set up another jam jar with the same quantities as jar one but put this jam jar in a cold place such as the refrigerator.
What effect did the extra sugar have on the growth of yeast? What effect did a reduction in temperature have on the growth of yeast?

class of living organism that is neither plant nor animal. Both bacteria and fungi are easy to grow and look after in vast quantities. They are also relatively simple to study. As a result, these two microorganisms have been used to help produce a wide range of products.

Enzymes

All organisms, including yeast and bacteria, contain a group of chemicals called enzymes. Enzymes are biological catalysts. A catalyst is a substance that will start or increase the rate at which a chemical reaction takes place but does not take part in the reaction itself. The role of enzymes in cells is to make sure that the cellular reactions take place fast enough to sustain life. For example, cells contain enzymes that control respiration, the process in which energy is released from food substances such as glucose. There are many thousands of different enzymes and each enzyme can work on only one particular reaction or group of reactions. The enzyme is said to be "specific" to a particular reaction. The remarkable thing about an enzyme is that, while it starts or speeds up the reaction, it does not actually take part in it, so the enzyme can be used over and over again.

An example of an enzyme is salivary amylase, which is produced by salivary glands in the mouth. The salivary amylase speeds up the breakdown of starch into a smaller substance called

In what ways are biological detergents better than non-biological detergents for the environment?

 Why do living organisms need enzymes?

 What is the best temperature for using biological detergents?

maltose, a type of sugar. This reaction, which takes just seconds in the presence of an enzyme, would take many hours without it. You can see for yourself how quickly the enzyme works. Simply chew a piece of white bread for a few minutes. As you continue to chew, you should notice that the taste changes; you will detect a sweet taste. This is due to the breakdown of starch into maltose.

Enzymes work best at the body temperature of the organisms in which they are found. For example, enzymes in human beings work best at 37°C (98.6°F), which is our normal body temperature. If the temperature increases, the structure of the enzyme is altered and it cannot work properly. If somebody suffers from a very high fever or from sun stroke, the enzymes can actually be destroyed. Enzymes are also affected by low temperatures. However, rather than being destroyed by the low temperature, they are merely deactivated. When the temperature rises again, they are able to work normally. Enzymes are required only in very small amounts, because they can be used over and over again.

Scientists are now using enzymes in industrial processes. Many detergents are now described as biological. This simply

EXPERIMENT

Investigating detergents

This experiment looks at the effectiveness of different types of detergents and compares the cleaning ability of biological detergents that contain enzymes with the non-biological types. You will need an old piece of white cotton material such as part of an old sheet, a selection of biological and non-biological detergents, a pair of scissors, four different types of food to stain the clothing (e.g., tomato ketchup, chocolate sauce, coffee and jam), a large container and a thermometer.

1 Cut the cotton into squares of equal size, each approximately 20 cm (8") square. You will need one piece of cotton for each type of detergent plus one extra.

2 Stain the cotton by taking a small amount of one type of food and smearing it over one-quarter of each of the cotton pieces. Wipe off any excess. Repeat this with the other three food samples, using a different quarter for each. Leave the cotton for a couple of hours to allow the stains to dry.

3 Fill the container with warm water at approximately 40°C (104°F) and add a small amount of one of the detergents. Mix the detergent in thoroughly and place one of the samples of cotton in the water. Wash the cotton by hand for five minutes and leave to soak for another five minutes. Then remove it, wringing out the water, and leave it to dry on the side.

4 Repeat the experiment using another type of detergent and another sample of cotton. Also repeat the experiment using just water (no detergent) with the extra piece of cotton. This will show you how much of the stain will come out using just warm water. This is a "control" and can be used to compare the effectiveness of the different detergents.

Which detergents worked most effectively? Were there any differences between how well the stains were removed?

TECHNOLOGY IN ACTION 27

The thick carpet of green algae in this California harbor has been caused by the combined effects of excess nutrients in the water and the warm weather.

The holes in this piece of wood have been made by the shipworm, a mollusk that can severely damage the timbers of ships. It can digest wood using a particular enzyme that scientists are now finding very useful.

means that they contain enzymes that make the cleaning process more effective at low temperatures. These enzymes are usually made artificially. The best temperature to wash with biological detergents is between 35°C and 40°C (87°–104°F). If the temperature of the wash is too high, the benefits of using a biological detergent will be lost. Most of the enzymes in detergents are those that will digest carbohydrates (starch and sugar) and fats. Others can digest proteins, so these enzymes can digest food stains, bodily secretions, such as sweat, and even grass stains. A mixture of enzymes can therefore remove most of the stains that appear on our clothes.

In recent years, sodium polyphosphate has been used to prevent some forms of dirt from sticking to the fabrics being washed. However, this increases the amount of phosphates in the water supply and contributes to water pollution, causing effects such as algal bloom in our rivers and ponds. With the increase in environmental awareness, detergent manufacturers have had to redesign their products. People have demanded new designs that are just as effective in cleaning clothes but contain reduced amounts of phosphate. In effect, one could argue that the new environmentally friendly detergents are a result of "natural selection." They are certainly much better suited to the environment.

Designing a new detergent is not easy. The manufacturers have solved the phosphate pollution but in doing so they have created new problems. The phosphate replacement in the detergent makes it more alkaline than before, reducing the effectiveness of many enzymes. So researchers have had to look for new enzymes that work efficiently in phosphate-free detergents. One enzyme that has potential in this area has been found in a bacterium living in the shipworm, a type of mollusk related to the clam, scallop and oyster. The shipworm can bore into the woodwork of ships by rotating its two shells. As it does so, it passes the wood into its digestive system. Strangely enough, the shipworm cannot itself

28 DESIGNS IN SCIENCE

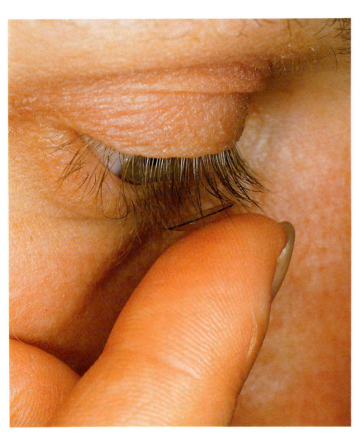

Enzymes from the shipworm may be used to make new cleaning solutions for contact lenses.

produce enzymes to digest the wood. Instead it relies upon two different types of bacteria living within its gut to produce enzymes that attack the cellulose and proteins in wood. Scientists have discovered that one of the enzymes works over a wide range of temperatures, being equally effective between 25°C and 50°C (77°–122°F). This enzyme may, therefore, be ideal for use in the new energy-efficient cold water detergents. And the enzyme appears to be unaffected by the alkaline conditions produced by the new detergents.

Other uses for this enzyme have also been discovered. Protein deposits build up on the surface of contact lenses, and these have to be removed using a cleaning solution, but most enzymes are destroyed by the disinfectant in lens sterilizing solution. However, the enzyme from the shipworm has been found to work well in these conditions. In future, contact lens wearers may be able to clean and sterilize using just one solution. The enzyme may also be used to recover silver from photographic film or remove hair (made of protein) from hides that are to be used to make leather. Many plants yield protein-digesting enzymes too. The kiwi fruit, for

EXPERIMENT

Enzymes from kiwi fruits

Gelatin is a protein. When a block of it is dissolved in water the gelatin causes the liquid to set. Kiwis contain an enzyme that digests protein. In this experiment you will discover what happens when you place mashed kiwi in gelatin. You will need one kiwi, two blocks of gelatin, a mortar and pestle, a spoon, a kettle, a knife, a potato peeler, a measuring cup and two small bowls in which to make the gelatin.

1 Carefully peel the skin from the kiwi. Cut the fruit in half and mash it.
2 Measure 500 cm³ (about 17 oz.) of water, place it in a kettle and bring the water to a boil.
3 Place a block of gelatin in each of the bowls. Pour half the boiled water into one bowl and the other half into the second bowl. Stir the liquid until the gelatin has dissolved.

4 Place the mashed kiwi into one of the bowls and stir the mixture well.
5 When cool, place both bowls of gelatin in a refrigerator to set.
After a couple of hours look at the gelatins. Have they both set?

TECHNOLOGY IN ACTION

 Protein-digesting enzymes extracted from the pineapple are often used in hospitals to remove dead skin before a skin graft is applied.

example, contains a protein-digesting enzyme that could be used to make meat tender.

Enzymes are used to make food products as well. One good example is the manufacture of cheese. To make cheese, milk has to be heated gently and made to clot. This causes the milk to separate out into a liquid called whey and solids called curds. First the whey is removed and the curds pressed together to form a cheese. This process of separation can be speeded up by using an enzyme called rennin. Young animals produce this enzyme in their stomachs so that they can digest the proteins in their mother's milk. However, the only way to extract the enzyme is to kill the animal.

Today scientists have been able to find yeast cells that make a similar enzyme that performs the same function. You can be sure that cheeses labeled "vegetarian" are those that have been made using the rennin substitute produced by yeast, rather than rennin from calves.

The first stage of cheese-making involves the separation of the solid curds from the liquid whey.

Biosensors

Enzymes have the ability to recognize one type of molecule in a mixture of many different molecules, even molecules that are very similar. They can also detect the molecules when they are in very low concentrations. For these reasons, scientists are using enzymes as biosensors. In a biosensor, the enzyme detects the target molecule and reacts with it. This creates a change in the mixture that can be monitored. Sometimes the change causes an electrical signal or a color change. Biosensors enable doctors to carry out very simple tests to diagnose disease. A very common example is the sugar test strip. This is a plastic strip with a pad of pink paper at one end. The pad contains an enzyme that can detect glucose. When the test strip is placed in a solution containing glucose, the pink color changes to a dark red within a few seconds. This is a very simple way of testing for glucose in the urine, a possible symptom of sugar diabetes. Similar test strips can be used to test for protein.

A diabetic tests the level of sugar in her blood using a biosensor.

30 DESIGNS IN SCIENCE

EXPERIMENT

Testing for sugar

In this experiment you discover how a biosensor works. You will use some test strips to check the amount of sugar in a liquid. You will need some sugar test strips (this can be obtained from pharmacists), sugar, four small jam jars, a teaspoon and water.

1 Fill all the jars with water from the tap.
2 Add half a teaspoon of sugar to the first jar. Stir the water until the sugar dissolves.
3 Add one teaspoon to the second jar and stir as before.
4 Add two teaspoons of sugar to the third jar and stir.
5 Add nothing to the fourth jar.
6 Test strips usually come with a color scale. When the test strip come into contact with sugar it changes color. The color indicates how much sugar is present. Take a test strip and place the colored end in the water of the first jar. Leave for a few seconds and remove. Look for any color change. Compare the color of your test strip with the color scale.
7 Repeat with the other three jars, using a fresh test strip each time. Does the color alter? Can you estimate how much sugar is present in each of the jars?

Cleaning up pollution

A compost heap relies on bacteria and fungi to break down the garden refuse.

Over the last 100 years or so, scientists have made a huge range of synthetic chemicals. Many of these chemicals were deliberately designed to be resistant to breakdown by organisms such as bacteria. For example, DDT (see pages 32–33) was designed to kill insect pests of crops as well as the malarial mosquito, while the chemical PCB was used as a coolant in electricity sub-stations. Unfortunately, these chemicals were found to have significant environmental side-effects. Since they are almost impossible to break down under normal conditions, these chemicals remain in the environment for many years. In the past, people were unaware of the long-term dangers, and waste containing hazardous materials was simply buried in the ground. The discovery of new strains of bacteria may now help us to clean up the environment.

Bacteria can be described as "nature's garbage men," naturally recycling waste. Nearly everywhere there are natural populations of bacteria decomposing waste material. In the

TECHNOLOGY IN ACTION | 31

Straw is often used to mop up oil spills. It has been found to be more effective than synthetic chemicals that are sprayed on oil to make it disperse.

 There are more than 30,000 hazardous waste sites in the U.S. alone that could be cleaned up using bacteria.

 Bacteria may even be used to clean up radioactive uranium from nuclear weapons.

Key words
Biotechnology the industrial use of living organisms to make food, medicines, etc.
Enzyme a biological catalyst that starts or increases the rate of reactions in living organisms.

garden it is bacteria that help rot down the compost heap. Many industries also make use of bacteria to treat their waste products. For example, a sewage treatment plant uses bacteria to digest sewage and render it harmless. Industrial processes such as paper-making and food processing use bacteria to clean their waste water.

Bacteria will soon be used to clean up sites where there are hazardous waste materials under the ground. At present, the only way to treat such a site is to dig up all the polluted soil and then destroy the pollutants by chemical treatment or burning, but this is very expensive. Recent trials have involved treating the pollutants on-site. A mix of water, oxygen and nutrients is injected into the soil. The oxygen and nutrients stimulate any naturally occurring bacteria in the site, thus speeding up their digestion processes. The water seeps through the site, picking up toxic chemicals, before being pumped out and fed into huge vats containing bacteria. In the vats, the bacteria break down the toxic wastes into carbon dioxide and water, which are both harmless. The water is pumped back into the soil. Each time the water recirculates, it gets cleaner.

Oil pollutes both land and water. We use so much oil in our everyday lives that huge amounts of crude oil have to be transported around the world in supertankers. The largest of these can carry over 500,000 metric tons (550,000 tons) of oil. Accidents inevitably happen, and there are occasional major oil spills at sea, which attract worldwide attention. But, just as important, small amounts of oil find their way into our rivers, down our drains and on to our soil from many different sources every day. Scientists are now experimenting with oil-eating bacteria. These bacteria were collected from places in the world where oil fields are close to the earth's surface, and oil seeps naturally out of the ground. The bacteria that live at these sites have evolved an ability to digest the oil. The bacteria are grown in the laboratory and then dried and stored as a powder. The powder can be spread over a polluted site either by hand or by machine. As yet, the oil-eating bacteria have been tried only in test sites, but the results look very promising.

Defense mechanisms

Most living organisms have to be able to defend themselves against attack by predators and disease-causing organisms. Any defense mechanism that gives an animal or plant an advantage over another competing organism will help that animal or plant to survive. As a result, evolution has resulted in a vast array of defense mechanisms.

The battle between attacker and defender is never-ending. Many of the most important battles today are those waged at microscopic scale, against the bacteria and viruses that cause disease. Living organisms, such as plants, fungi and insects, have naturally evolved a huge range of substances to protect themselves against other living organisms, and even to kill them. Medical researchers, in a similar way, are developing a widerange of substances to help us fight disease, while agricultural researchers are working on chemicals that can guard against the many pests that affect our crops and livestock.

The ladybug is brightly colored to show that it is poisonous to birds. This is its defense mechanism.

Pesticides

A pest is an organism such as an insect, fungus or rodent that has a harmful effect on human beings. Most pests damage crops, but they may also harm livestock, damage buildings or invade homes. To prevent infestations of pests, or deal with them once they have occurred, people have produced chemicals that will kill the pests. There are three main groups of pesticides, named after the type of organism that they harm. Insecticides kill insects, fungicides attack fungi and herbicides are used to kill weeds.

There are many different types of insecticides, but most work on the principle of being absorbed into the body of the insect when the insect comes into contact with the chemical. A well known example is DDT, an insecticide that was used extensively during the 1950s and 1960s to kill insect pests. Unfortunately, DDT did not break down naturally in the environment, for it is not biodegradable. Instead, it persisted at low levels for

The locust (right) is one of the many pests that have to be sprayed with pesticides in order to control their numbers (above).

TECHNOLOGY IN ACTION 33

many years. When animals ate food contaminated by DDT, they accumulated the pesticide in the fatty tissue of their own bodies. The animals at the top of food chains, the carnivores, were most affected by DDT. It was a particular problem for birds. Over a number of years, the amount of DDT would build up as more contaminated food was eaten. Over winter, when there was a general lack of food, their bodies would start to use the fat store and in doing so a lot of DDT would be released into the bloodstream at one time, frequently killing the bird. Another side effect of DDT poisoning was that female birds would lay eggs with very thin shells. The eggs would often break during incubation, and the developing bird would die. As a direct result of the use of DDT by farmers, the populations of predatory birds fell.

Today, DDT is banned in most countries, although it is still used in a few places to control the malarial mosquito. It is cheap to produce and ideal to spray onto the ponds and the surfaces of buildings where mosquitoes and their larvae can be found. Because malaria kills millions of people

The number of peregrine falcons plummeted as DDT spraying increased.

 DDT has been so effective against malaria in Mauritius that the annual infant mortality rate has fallen in 10 years from 150 infant deaths per thousand to just 50.

 Can you think of five pests that live around your home?

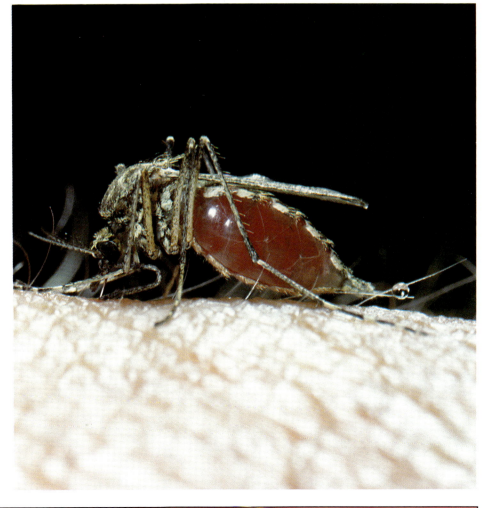

DDT is one of the most effective pesticides against the malarial mosquito.

34 DESIGNS IN SCIENCE

DDT has even been found in the livers of penguins living in the Antarctic, thousands of miles away from the nearest place where DDT was used.

each year, it has been decided that in places where malaria is a problem the beneficial effect of spraying DDT outweighs the environmental side effects.

In recent years, researchers have developed more powerful chemicals that do not stay in the environment, to replace DDT. Natural insecticides, such as derris and pyrethrum, can be prepared from plants but, although they are safer to use, they remain effective for a much shorter time, so they have to be applied more often.

Natural pesticides

? *Why do plants need to protect themselves with natural pesticides?*

As pests gradually develop resistance to artificial pesticides, researchers are looking at the natural chemical pesticides produced by plants. One plant that has attracted attention is the neem tree that grows in the tropics. It produces a natural pesticide that deters more than 100 different types of insect from feeding on the tree. Some companies have incorporated extracts of the neem tree in new pesticides. One product protects against the most common insect pests, such as whiteflies, while having the benefit of being harmless to humans. Another part of the neem tree's

 The tobacco plant can produce a substance that will actually dissolve the body of an insect.

The cockroach can be controlled using a naturally occurring fungus. The cockroach picks up the fungus, thinking that it is food, and carries it back to its nest where the fungus grows and attacks the cockroaches.

TECHNOLOGY IN ACTION 35

defenses is a chemical that kills the insect as it changes from a larva to an adult.

Other plants produce similar defensive chemicals. The bark of the paw paw produces a substance that kills both juvenile and adult insects by preventing the insect from obtaining energy from sugar.

These natural pesticides break down very quickly, which is good for the environment, but, if a farmer were to spray such pesticides on a crop, it would be necessary to respray every few days. Genetics, however, may have an answer. Scientists are trying to insert the genes for making the various chemicals into bacteria that can survive for longer periods. They may even be able to actually insert the gene into the crop plant itself.

The medicine tree found in the Amazon is one of the many trees that can produce substances that could control pests and provide cures for many human diseases.

Antibiotics

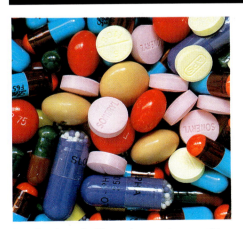

A collection of pills and capsules used to treat disease and medical conditions. Many pills are now developed from natural sources.

A wide range of chemicals are used as medicines to treat diseases. Many of these chemicals originate in the natural world. For a very long time people used certain remedies without really knowing how the chemical actually worked. But, as our understanding and technology improves, we are becoming better at designing new medicines and improving existing ones. Although many types of bacteria are useful and even essential to our survival there are some types that cause disease in people, for example cholera and dysentery. Fortunately, certain fungi produce chemicals that can be used to kill harmful bacteria. These chemicals are called antibiotics. The best-known antibiotic is penicillin, which is produced by a mold called *Penicillium*. The antibiotic was first discovered by Alexander Fleming in 1928 when he observed that the fungus had slowed down the growth of bacteria. However, it was many years before penicillin was identified and isolated. It

36 DESIGNS IN SCIENCE

The antibiotic penicillin is obtained from a mold called *Penicillium*.

 More than 14 percent of all reported cases of tuberculosis in the U.S. are resistant to one or more of the antibiotics used to treat the disease.

The skin of the clawed toad produces its own antibiotic, protecting the toad against disease.

was first tested on people in the early 1940s when it was used to treat war wounds and many lives were saved. Since that time, thousands of antibiotics have been discovered, though only a few are used in medicines.

There are two main groups of antibiotics. There are broad spectrum antibiotics, such as tetracycline and chloramphenicol, which are effective against a wide range of bacteria. The other group is called narrow spectrum since they can be used against only a few types of bacteria. These have more limited uses. Unfortunately, extensive or uncontrolled use of an antibiotic, particularly the broad spectrum antibiotics, can lead to bacteria becoming resistant to the chemical. Since bacteria reproduce very quickly, any change in the DNA (see pages 21–22) of the bacteria quickly spreads throughout the population. Widespread use of streptomycin, which acts against bacteria that cause blood infections, has produced resistance in more than half of the known strains of such bacteria. Sometimes doctors use combinations of two or more antibiotics to combat a disease. Each antibiotic works in a slightly different way, so it is less likely that the bacteria can resist the effect of such an antibiotic "cocktail."

Scientists have recently discovered antibiotics that are produced naturally by animals such as the frog. These natural antibiotics may be more powerful than existing ones, for they seem to work equally well against bacteria, protozoa and fungi, and some may even act against cancers. For example, the African clawed toad produces an antibiotic in its skin that makes it resistant to infection by any microorganisms in the water. The antibiotic can be extracted from the frog by simply rubbing its skin; the antibiotic seeps out in the form of a milky white fluid. These toads secrete the same fluid if they are scared, perhaps as a means of protection.

Sharks, too, produce a powerful natural antibiotic. They produce a chemical called squalamine. Squalamine belongs to a group of chemicals called steroids. Many human hormones, including reproductive hormones, are also steroids. Therefore scientists believe that it will be quite easy to adapt squalamine as a medicine for humans. Furthermore, it has been discovered that sharks do not suffer from cancers, and it is possible that this is due to the presence of squalamine.

Research has shown that animal antibiotics work differently from penicillin. They attack bacteria, but do not destroy them directly.

TECHNOLOGY IN ACTION 37

An ancient witch's brew!

This quote, from Shakespeare's *Macbeth*, describes what sounds like a revolting mixture of ingredients. But could it have been medically effective?

> Eye of newt, toe of frog,
> Wool of bat, and tongue of dog,
> Adder's fork, and blind-worm's sting,
> Lizard's leg, and howlet's wing,
> For a charm of powerful trouble,
> Like a hell-broth boil and bubble.

They rip open the wall of the bacterium, allowing the cell contents to pour out and be attacked by the normal immune system. This technique means that the bacteria cannot become resistant to the antibiotic by simply mutating. New medicines based on these animal antibiotics are still at the testing stage, but we may well see them used to treat impetigo (a serious skin disease), eye and stomach infections, and some cancers.

For what types of illness does a doctor prescribe penicillin?

Key words
Antibiotic a chemical that kills bacteria and fungi.
Pesticide a chemical used to kill pests such as insects, fungi and weeds.
Protozoa simple single-celled animals such as the ameoba.

Scarlet macaws in the Amazon swallow kaolin, a type of clay, as an antidote to the poisonous berries that they eat. Humans use kaolin to treat upset stomachs.

38 DESIGNS IN SCIENCE

Control systems

Animals and plants are complex organisms, consisting of many interrelated systems. The many different organs and processes all need to be monitored and controlled. Machines are also complex and they too require control mechanisms.

Perhaps surprisingly, the control systems of living organisms and artificial structures operate on very similar principles. They both have sets of sensors to provide information about the outside world, effectors to bring about a change, a communication system to pass messages and a processor to work out what to do. This is as true in the animal kingdom as it is for any machine.

A natural control center

Higher animals such as the mammals have a well-developed nervous system consisting of a large brain and a spinal cord, with nerves connecting them to the rest of the body. These nerves transmit information to and from the central nervous system. A nerve is, in fact, a bundle of specialized cells contained within a protective sheath. It is just like an electrical cable, which consists of a tough outer covering within which are up to three different wires, each having its own thinner covering. The specialized cells are called neurons. Each cell has a long thin thread, which, when stimulated, sends information along its length in the form of electrical impulses. There are several types of neuron: sensory, motor, and associative. Each varies in structure and function.

A sensory neuron carries information such as temperature, pain or pressure to the central nervous system. This information is carried on to the brain, but it may trigger an immediate response to be sent back by a motor neuron. The motor neurons carry messages to the muscles, causing them to relax or contract. If your finger touches a hot object, a pain message is sent along the sensory neuron running from the skin, along the arm to the spinal cord. Here the message is picked up by an associative neuron that sends information to the brain. At the same time, it sends another message along a motor neuron to the muscle in the arm, causing the muscle to contract, and so pulling the finger away from the hot object. Although the brain is aware that this has happened, it has not consciously told the arm to move. This type of instant

A close-up of a human motor neuron, which sends messages to a muscle causing it to contract.

Impulses travel along human neurons at up to 160 meters (176 yd.) per second.

TECHNOLOGY IN ACTION 39

Sneezing is an automatic reflex.

reaction is an example of nerverreflex; an automatic action that is carried out without thinking. It is usually protective, and very rapid. Other reflexes include contraction of the pupil in the eye, sneezing and blinking.

The brain is involved with many activities that require simultaneous actions. For example, the brain must cope with information from an image falling on the retina of the eye, sound reaching the ear, as well as touch, smell and taste. All of this data is carried by nerves to special parts of the brain where the information is received and interpreted.

The human brain is divided into the front, middle and rear areas. The front brain is concerned with higher thought processes and the sensory centers. At the very front is a region concerned with personality and memory. The central region is where visual and aural sensory information is received. Other areas control the transmission of messages to the muscles to control movement. The rear brain is concerned with controlling the main bodily processes over which we have no direct control. The medulla controls the rate at which the heart beats and the lungs inhale, while the cerebellum maintains coordination and balance. We can, however, override some functions of the rear brain. For example, we can control the rate at which we breathe.

 What might happen if a sensory neuron was damaged?

EXPERIMENT

Reaction times

In this experiment you will discover just how quick your reactions can be. You will need a meter ruler or yard stick or a thin pole of wood of about the same length, a notebook, pencil and a friend to help.

1 Ask your friend to hold the ruler so that it is positioned vertically just above your hand (see the photograph). Hold your hand open so you are ready to catch the ruler when your friend releases it.

2 When your friend is ready tell him or her to let go of the ruler, making sure that the ruler falls straight. As the ruler falls, you have to try and catch it as it drops between your fingers.

3 Make a note of how far the ruler dropped. You may want to try this a few times and work out an average value. The further the ruler drops, the slower are your reactions.

You could repeat the experiment on your friend and compare the results. You could also try this using your other hand. Is your reaction time the same with both hands?

DESIGNS IN SCIENCE

Microchips and computers

Machines are now becoming far more complex and able to carry out more functions due to developments in electronics. Electronic engineers fit complex electric circuits onto tiny pieces of silicon, often called microchips. Almost all modern machinery is controlled by one or more microchips, and the latest computers (or microprocessors) consist of little else but connected microchips, each with its own function. Although microprocessors can be used to control quite complex processes, each chip has to be programmed before it can work. The program is a set of rules that enable the microchip to function correctly.

Just as human beings have a brain in which information is stored and processed, so a computer has a memory and a processor. And in just the same way that a human body has a nervous system to pass messages from the brain around the body (see page 38), so a computer has a data conductor around which its messages are passed. Although we may think of computers as being very much faster and more accurate than human beings, this is not the case. Most computers are unable to carry out tasks for which they have not been programmed. So they are not intelligent.

Some of the latest computer-controlled machines are helping people to overcome disabilities. For example, microchips are being tested for their ability to stimulate nerves directly. This would mean that sensory information could be sent to the brain of people who were profoundly deaf or totally blind.

Computers are being used to provide a model, in software, of real-world systems or structures. These simulations allow scientists and engineers to repeatedly test designs in different situations, or to compare different designs, and to solve problems of performance before building anything at all. Computer models can even show dress designers what a new fabric will look like when cut and worn as a dress, before the cloth has even been woven, let alone dyed. The computer would be able to provide the designer with a blueprint.

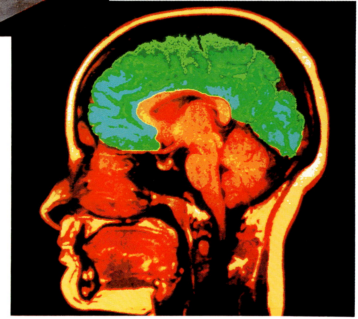

The computer acts in the same way as the brain does in the human body. It contains a memory and information processing section, which enable it to carry out certain tasks.

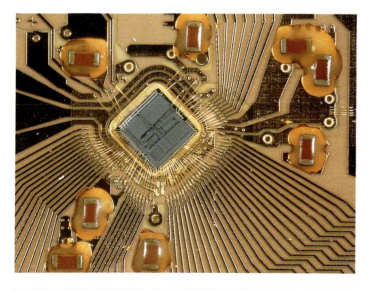

The microchip can be found in all kinds of modern equipment. Computers are made up of lots of linked microchips.

TECHNOLOGY IN ACTION | 41

Feedback mechanisms

> ! *The very first computer was designed by Charles Babbage in 1835, but it was too complex for the technology of the day, and it was never completed.*

Have you ever wondered how our body temperature stays at 37°C (98.6°F) regardless of the temperature of the external environment? Our bodies have a sort of internal thermostat that regulates our temperature. If our body temperature rises above normal, we start to sweat and heat energy is removed from our skin by the evaporation of the water. This helps return the body to its normal temperature. If we get cold, we shiver. This produces heat energy, which warms the body up. Any change away from the normal temperature starts a process that brings the temperature back to normal. This is called a negative feedback mechanism.

An air conditioning and heating system in a building has been designed to work in much the same way. If the air temperature falls below a set value, a sensor triggers the heating system to switch on and this heats the building up. If the temperature rises above the required temperature, the system responds by chilling the air. The computer that controls the building's air conditioning is working in just the same way as the human brain controls the body temperature. The microprocessor receives the temperature information, compares it to the required temperature that it has stored in its memory, and brings about an action to change the heat flow if necessary.

In this paper mill the quality and thickness of paper is measured by an optical sensor that sends information back to the control unit.

Modern industries make extensive use of these feedback systems to control automated production lines. In paper-making, optical sensors are used to check the thickness of the paper and to see if the watermark is in the right position. The computer is provided with information concerning the ideal thickness of the paper, and it compares the reading from the sensors with that of the memory. If there are any differences, the computer changes the pressure of the rollers earlier in the paper-making process to correct the error. The continual checking of the paper, several hundred times every second, ensures a top quality paper with almost no variation in thickness.

The process of feedback is very important in both natural and artificial mechanisms. It provides a continual check on systems, making sure they are working properly.

 Can you name three devices in your home that store information?

42 DESIGNS IN SCIENCE

Artificial intelligence

Kayaking requires the coordination of different parts of the body. Information from the eyes is sent along sensory neurons to the brain and then a message is sent along a motor neuron to the muscles in the arm.

Artificial nerve cells may be one million times quicker than natural ones.

The largest neural network computers use several thousand artificial neurons, but the human brain has billions of neurons.

The human brain is very complex indeed. In a simple situation, such as a person dining in a restaurant, that person is capable of carrying on a conversation, while being able to control the body's movement and monitor input from eyes, ears, nose, tongue and fingers. Computer designers are hoping that, one day, they will be able to produce a computer system that can compete with the human brain. However, most of the current computer systems that are classified by the term "artificial intelligence" are actually not intelligent at all. Many are more accurately known as "expert systems." This type of computer has a set of "rules" created by a human expert, which it applies to a particular problem. These computers have proved very capable in such fields as medical diagnosis because they can be programmed with a large amount of knowledge from the best experts in the field.

The latest generation of artificial intelligence machines is based around a system called the neural network, which copies the structure of the human brain, which is a highly interconnected network of billions of individual brain cells, or neurons. Such systems can be "trained" and can "learn" from their mistakes. However, while such systems are undoubtedly very impressive, they cannot yet be classed as intelligent, for they do not demonstrate any understanding of what they are doing.

A new silicon neuron that behaves just like a natural nerve cell has been developed. It is so small that 200 of these artificial cells would fit on a silicon chip just one centimeter square in area (1.16 sq. in.). This is, perhaps, a small step toward the development of a complete artificial nervous system. In a natural neuron, the message is in the form of an electrical signal that travels rapidly along the length of the cell. The artificial silicon nerve cells create a flow of electrons that mimics the electrical activity of a nerve cell. In the future, many such silicon chips may be linked up to create an artificial brain.

TECHNOLOGY IN ACTION 43

Robots

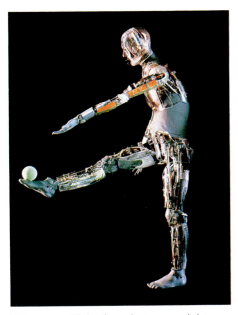

When you think of a robot, you might envision something more like this. Unfortunately the intelligent robot is far from being made.

In the 1970s, it was predicted that, by the year 2000, robots would be a widespread and familiar part of everyday life. The idea then was that the robot would be a copy of a human being but without any human weaknesses. This now seems very unlikely to happen for many, many years, for most robots are still extremely simple when compared with the complexity of living creatures. Almost all modern robots are used in static locations for repetitive work such as making things on an assembly line.

Industrial robots normally consist of a number of jointed arms with a manipulator at the end. The position of each arm can be controlled by a microprocessor, and sensors provide positioning information. Feedback systems are often used to help the robot keep the right position. The "muscles" of a robot are usually provided by motors or, for more power, hydraulic pumps. These robots are strong, accurate, do not get tired and are quite versatile. However, they do require careful programming. Most industrial robots are fixed to the floor, but some are mobile, usually following buried signal cables so that they do not stray from a fixed route. Because of the lack of real progress in artificial intelligence and image recognition, both of which are very complex subjects requiring enormous computing power, robots cannot yet make enough sense of their surroundings to be allowed to move freely about an unfamiliar environment.

The latest designs of robot are to be used to help surgeons to perform hip replacements. Trials have found that a robot can drill a hole for the hip joint replacement more accurately than a surgeon can.

A robotic arm is used for accurate welding. This is a simple repetitive job that is best suited for an easily programmed robot.

Key words
Microchip a miniature electronic circuit on a chip of silicon.
Neuron a specialized cell capable of transmitting impulses.
Nerve a bundle of neurons carrying information to and from the central nervous system.

The future

Much of the development of biotechnology and artificial life will involve the use of very powerful computers. The latest computers can be used to help create new molecules. Scientists can create these so-called designer molecules from scratch, using their existing knowledge of chemistry. The use of a computer enables them to assemble the molecules quickly and easily. The computer can show the position of every single atom in the new molecule in three dimensions. It is even possible to carry out tests to see how the molecule will react with other molecules. Such tests now take hours, rather than the months or years taken using conventional methods. In the future, "designer molecules" may be able to fight disease, for the next generation of these creations will be designed to behave in a similar way to antibodies, the body's own means of defense (see pages 35–37).

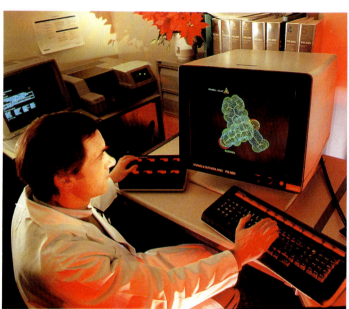

Computers are used to design complex structures, such as molecules.

Not only can scientists now fight disease, they can also deliberately alter the pattern of evolution of other living organisms. Over the last few hundred years, people have altered the characteristics of species by artificial selection. Our understanding of the biology of the mammal is now so complete that cloning (the production of genetically identical organisms) animals, will soon become commonplace. Cloning is very similar to the natural process that produces identical twins.

Bacteria are not the only living organisms that may be used in the future to clean up toxic wastes (see pages 30–31). Plants such as ragweed have been found to accumulate toxic metals in their cells. At the end of the year, these plants can be harvested and burnt in an incinerator. It may even be possible to reclaim the metals.

TECHNOLOGY IN ACTION 45

Identical twins develop when a fertilized egg divides into two and each new cell develops into a separate individual, both individuals having identical genes. In the future, it will be possible to take an embryo from a prizewinning cow, for example, and clone it to produce many identical embryos. The embryo would then be placed into the uterus of another female cow, rather like a test-tube baby. In 1993, the first human embryos were cloned. American scientists took 17 human embryos and split them to form 48 embryos. The embryos did not live for very long. There is, however, considerable debate over the morality of this kind of experimentation.

Another form of research into the makeup of human beings is being carried out in the U.S. and in other parts of the world. Scientists are trying to piece together the genetic code of human DNA. They may eventually be able to identify all the individual genes on a chromosome. People can already find out whether they are carrying genes associated with certain diseases. In the future this will be possible for even more diseases. This knowledge can be a powerful way of preventing many serious illnesses. For example, the information can help doctors to predict the likelihood of a baby being born with an inherited disease and hence provide advice to prospective parents. Again, there are many moral issues still to be resolved in this area.

A scientist looking at a cell from a patient suffering from a genetically inherited disease. The genes are closely studied to see which ones caused the illness.

The importance of studying living systems has given rise to a new area of research called biomimetics, or biological mimicking. This involves using biological information to design new materials and manufacturing processes. Scientists are trying to unlock the secrets of many natural designs: for example, how rats can gnaw through metal without wearing out their teeth, how walnut and coconut shells resist enormous forces without cracking, and why beetle shells are as tough as tank armor. Biological processes have been tried and tested by evolution over billions of years and have provided solutions to some of the most complex problems. We are only just discovering some of these solutions, and there are many more answers as yet undiscovered.

Glossary

adaptation the process of adjustment of an organism to environmental conditions.
antibiotic a drug that kills or inhibits the growth of harmful bacteria and fungi.
bacteria single-celled microorganisms.
biodegradable capable of being broken down by living organisms such as bacteria and fungi.
biotechnology the industrial use of living organisms to make things such as food and medicines.
environment the surroundings of a living organism together with all the living and non-living factors that may affect its survival.
enzyme a biological catalyst that starts or increases the rate of reactions in living organisms.
evolution the gradual change in the characteristics of an organism.
fungus a living organism that is neither animal nor plant; often a parasite.
genetic engineering the alteration of the genetic composition of an organism.
hormone a chemical messenger, produced by special glands, that controls the various bodily activities.

immune system the natural defense system that protects the body against infection by disease-causing organisms.
mammal an animal in which the female gives birth to live young and produces milk.
microchip a miniature electronic circuit on a chip of silicon.
mutation a sudden change in genetic information.
natural selection theory that those species best suited to an environment will live to reproduce, passing on their characteristics.
nerve a bundle of neurons, carrying impulses to and from the central nervous system.
neuron a specialized cell capable of transmitting electrical impulses.
pesticide a chemical used to kill pests such as insects, fungi and weeds.
prototype a trial version of a product that is tested extensively before manufacture begins.
protozoa simple single-celled animals such as the amoeba.
silicon a crystal used in the manufacture of microchips.
technology practical application of science in everyday life.
variation differences, a change from the normal.
virus an acellular parasite.

Answers to the questions

p. 9 There are many different features – position, shape and color of eyes, skin color, shape of face, nose, lips, presence of freckles.

p.13 This is an open question so there is no right or wrong answer – some suggestions might be appropriate here.
Features like – color, speed, space, number of seats, audio system, electric windows, security system, central locking, luggage storage, air conditioning.
Features to change – could be any of the above.
Changes – larger cars, more space, bigger engine so faster etc.

p.17 Plants can be pollinated by insects or birds, even bats, which carry pollen from one flower to another. Wind can carry pollen between flowers and some aquatic plants rely on the water.

p.19 The rain forests affect the world climate, the water cycle, approx. 50 percent of all the species live in the rain forests, some of these species could be sources of new foods and medicines. Good quality timber such as teak and mahogany are found in these forests. Some of the plant species could be used in future plant breeding programs to improve our crop plants. Rain forests hold a lot of water – rather like a sponge – and this prevents flooding in places downstream.

p.22 If scientists use genetically altered bacteria carefully, it should be safe to release these bacteria into the environment. However, there is a possibility that these modified bacteria may undergo change and evolve. There is the possibilty that they could become harmful.

p.23 4096. For example 1 bacteria doubles to 2 in 20 minutes, 2 become 4 in 40 minutes, 4 become 8 in 60 minutes, 8 become 16 in 80 minutes and so on.

p.24 Beer, bread and some cakes.

p.26 Biological detergents require a lower temperature to work and so save on heat energy. This in turn reduces the amount of fossil fuels burnt in power stations and reduces carbon dioxide emissions. Generally a smaller amount of the detergent is also used in the washing machine.

p.26 Enzymes are found in every living cell. They enable cell reactions to occur quickly. They can also control the rate at which the reactions take place. Enzymes are used to digest our food. They are used in cell respiration. Toxic waste products are detoxified by enzymes in the liver. Enzymes are even found in our tears. Plants use enzymes in photosynthesis. Without enzymes cell reactions would take place too slowly for life to exist.

p.26 Best temperature is approximately 35-40°C, which is close to our own body temperature.

p.33 Cockroaches, rats, mice, fleas, squirrels, flies such as mosquitoes and bluebottles, silver fish, termites, moths that damage clothes, beetles that damage timber.

p.34 Plants need to prevent herbivorous animals from eating their leaves, flowers, roots etc. If their leaves are eaten or damaged plants cannot photosynthesize so well.

p.37 Penicillin is prescribed for bacterial diseases and infections. Penicillin was used during the Second World War to treat people who had infected wounds. Now it is used against a wide range of bacterial infections.

p.39 Damage to a sensory neuron would mean that messages from the sense organ or receptor would not reach the central nervous system, therefore the person may not be able to feel anything, or in the case of damage to the optic nerve, not see anything.

p.41 Devices that store information include computers, programmable cameras, washing machines, dish washers, answerphones, security system, heating system.

Index

Key words appear in **bold face** type.

A
adaptation 9, 15, 46
African clawed toads 36
agriculture 17–18
air conditioning 41
algae 27
alkalinity 27–28
amylase *See* salivary amylase
animal breeding 20–21
antibiotics 35–37, 46
antibodies 44
appendix (part of the human body) 9
architects 14
artificial intelligence 42
avocets 11

B
Babbage, Charles 41
bacteria
  antibiotics used against 35–37
  biotechnology and 25–26
  definition of 23, 46
  in detergent production 27–28
  for environmental cleanups 30–31
  enzymes produced by 25
  in pesticide production 35
  reproduction of 22
  structure of 21–22
baking 24
beer 25
bees 11–12
beetles 45
bills (of birds) 11
biodegradable substances 23, 32, 46
biomimetics 45
Biopol 23
biosensors 29–32
biotechnology 24–31, 46
birds 11, 33
blueprints 14–15, 40
bones 11
brain 38–40, 42
bread 25–26
brewing 24
broad spectrum antibiotics 36

C
camouflage 6, 8
cancer 36–37
carbohydrates 27 *See also* starch; sugar
carbon dioxide 25, 31
carnivores 33
cars 13
catalysts 25
cells 14–15
cellulose 28
central nervous system 38
cereals 18
cerebellum 39
cheese 29
chloramphenicol 36
chromosomes 14–15
cloning 44–45
cockroaches 34
coconuts 12, 45
compost heaps 30
computers 15, 40–42, 44
contact lenses 28
control systems 38
cows
  breeding of 20
  genetic engineering of 23
  insulin from 22
cross-pollination 16–17
cytoplasm 15, 21

D
Darwin, Charles 6
DDT 30, 32–34
deforestation 19
derris 34
designer molecules 44
detergents 26–28
diabetes 22, 29
DNA 14, 21–22, 36
dogs 21
dress design 40

E
eggs 33
embryos 16, 45
environment 10, 14, 46
  See also pollution
enzymes 25–29, 46
evolution 6–9, 46
expert systems 42
eyes
  color 15
  structure 9

F
falcons *See* peregrine falcons
farming *See* agriculture
fats 20–21, 27, 33
feedback mechanisms 41, 43
fertilization 20
Fleming, Alexander 35
flowers 12, 16
fossils 6, 9
frogs 36
fungi
  antibiotics produced by 35
  for cockroach control 34
  in compost heaps 30
  defense mechanisms of 32
  definition of 24–25, 46
fungicides 32

G
gelatin 28
genes 14–15, 21
genetic code 45
genetic engineering 21–23, 46
giraffes 7
glucose 29
grasses 18

H
hair color 15
hands 10
hazardous wastes 31, 44
hearing aids 40
heart disease 20–21
hedgehogs 10
hemoglobin 23
herbicides 32
hip replacement surgery 43
hormones 46 *See also* insulin; steroids

I
identical twins 44–45
immune system 37, 46
industrial revolution 8
insecticides 32
insulin 21–22

K
kaolin 37
kayaking 42
kiwi fruit 28–29

L
ladybugs 32
landrovers 13
lizards 6
locusts 32

M
macaws *See* scarlet macaws
maize 19
malaria 33–34
maltose 26
mammals 38, 46
Mayan civilization 19
medicine trees 35
medulla 39
microchips 40, 43, 46
microprocessors 40
milk 20, 23 *See also* cheese
molecules 29, 44
mollusks *See* shipworms
mosquitoes 33
moths *See* peppered moths
motor neurons 38
muscles 38
mushrooms 12, 24
mustard plants 23
mutations
  in bacteria 36–37
  definition of 15, 46

N
narrow spectrum antibiotics 36
natural selection 6–7, 9–10, 46
neem trees 34–35
negative feedback 41
nerves 38, 43, 46
neural networks 42
neurons 38, 43, 46
nuclear weapons 31
nucleus (of a cell) 14–15, 21

O
oil (petroleum) 23, 31
oils, vegetable 23
optical sensors 41
orchids 17
oystercatchers 11

P
paper making 41
paw paw 35
PCB 30
pedigree breeds 21
penguins 34
penicillin 35–36
peppered moths 8
peregrine falcons 33
pesticides 32–35, 46
pests 32
petroleum *See* oil
PHB 23
phosphates 27
pigs
  breeding of 20
  insulin from 22

pineapples 29
pipelines 6–7
plant breeding 16–19
plastics 23
pollination 11–12, 16–17
pollution 8, 27, 30–31
potato plants 23
programming
 of computer chips 40
 of robots 43
proteins 27–29
prototypes 6–7, 46
protozoa 36, 37, 46
pyrethrum 34

R
racial differences 8
ragweed 44
rain forests 18–19
rats 45
reaction time 39
reflexes 38–39
rennin 29
reproduction
 by animals *See* animal
  breeding
 by bacteria 22
 by plants 12 *See also*
  plant breeding
robots 43

S
salivary amylase 25–26
scarlet macaws 37
seed banks 18–19
seeds 12–13, 16–17
selection 15–16 *See also*
 natural selection
self-pollination 16
sensory neurons 38
sewage treatment 31
sexual reproduction 17
sharks 36
sheep 23
shipworms 27–28
silicon 40, 42, 46
skin grafts 29
sodium polyphosphate 27
species 6–7, 21
sperm 20
spinal cord 38
spores 12
squalamine 36
squid 9

starch 25–26
steroids 36
straw 31
streptomycin 36
sugar
 as energy source in
  insects 35
 starch broken down into
  26
 testing for 29–30
sunflowers 19
supertankers 31
surgery *See* hip
 replacement surgery
survival 6, 10–11, 15
sycamore trees 12

T
technology 7, 46
test strips 29–30
tetracycline 36
thermostats 41
toads *See* African clawed
 toads
tobacco mosaic virus 22–23
tobacco plants 22–23, 34
tomato plants 18

tongue rolling 15
toxic wastes *See*
 hazardous wastes
trilobites 6
tuberculosis 36
tubes 11
twins *See* identical twins

U
uranium 31

V
variation 8–10, 15–16, 46
viruses 22–23, 46

W
walnuts 45
weeds 32
welding 43
wheat 12–13, 18
white flies 34
wine 25
wool 23
worms 6

Y
yeast 24–25, 29